# YOUR KNOWLEDGE HAS VALUE

- We will publish your bachelor's and master's thesis, essays and papers

- Your own eBook and book - sold worldwide in all relevant shops

- Earn money with each sale

Upload your text at www.GRIN.com
and publish for free

**Bibliographic information published by the German National Library:**

The German National Library lists this publication in the National Bibliography; detailed bibliographic data are available on the Internet at http://dnb.dnb.de .



**Imprint:**

Copyright © 2018 GRIN Verlag
Print and binding: Books on Demand GmbH, Norderstedt Germany
ISBN: 9783668931138

**This book at GRIN:**

https://www.grin.com/document/463908

Marvin Brucker


# Does Virtual Reality Gaming Evokes Stronger Brain Activities than Traditional Gaming?


## On the Usage of Virtual Reality


GRIN Verlag

**GRIN - Your knowledge has value**

Since its foundation in 1998, GRIN has specialized in publishing academic texts by students, college teachers and other academics as e-book and printed book. The website www.grin.com is an ideal platform for presenting term papers, final papers, scientific essays, dissertations and specialist books.

**Visit us on the internet:**

http://www.grin.com/

http://www.facebook.com/grincom

http://www.twitter.com/grin_com

# Virtual reality gaming evokes stronger brain activities than traditional gaming.

**Marvin Brucker**
Xi'An Jiaotong-Liverpool University
Suzhou

## ABSTRACT

Virtual reality influences an increasing number of areas in our daily entertainment life and reaches millions of people worldwide. While the underlying technology isn't completely new, its wide distribution as a consumer market device on the other hand is a new and ongoing trend. One of the main fields of application for the modern virtual reality devices is the gaming domain. But besides an incredibly high degree of realism, players also report about dizziness, exhaustion and intolerances of this new way of gaming. To what extend virtual reality gaming has a different impact on the gamer than traditional gaming is highly relevant for different target groups. This paper therefore evaluates whether virtual reality gaming evokes different brain activities than traditional gaming and how significant these differences are.

## Author Keywords

Virtual reality; gaming; brain activities.

## ACM Classification Keywords

I.3.7.[Computer Graphics]: Three-Dimensional Graphics and Realism - Virtual Reality; H.5.2 [Information interfaces and presentation]: User Interfaces - Graphical user interfaces.

## INTRODUCTION

### Background

Videogames in general, regardless of the platform and the genre, continue to increase in popularity [7]. The growing distribution of games and the accompanying increasing significance of games as a medium of entertainment created an own field of research, creating numerous studies showing the benefits [3, 6] and risks [5] of gaming. The growing popularity among users and therefore scientist is, inter alia, based on the development of games, which are nowadays able to deliver virtual worlds for the user that are very close to reality. One important factor to create this realistic user perception is often described by users as a feeling of "being

in the game". This feeling has been the object of many studies, trying to define and measure it. While the experience that a user during a media reception or the interaction with technology in general has, is usually referred to as user experience [2], the term user experience is primarily used in the human-computer-interaction field of research and is understood as an overall concept for all experiences with a system. The concept is based on various underlying concepts describing various aspects of the user experience. Absorption, Flow, Engagement, Immersion, Involvement and Presence are just some. The concept of presence though is the most researched one among these. This concept's high-level definition can be expressed as the impression of being there [1] and that presence describes the impression to experience a virtually transmitted environment as if it was real and not artificially transmitted by a system [11].

The usage of virtual reality headset, or head-mounted displays, is another approach of improving the perceived presence feeling within virtually created worlds. Just like videogames, virtual reality headsets are also enjoying a growing popularity, especially in recent year [9]. Virtual reality has also lead to an independent research area with numerous studies. Most of these virtual reality related studies take advantage of the high level of perceived presence, allowing researchers to place subjects into situations that would be too dangerous in real life. These studies show that with the usage of virtual reality headsets, the perceived level of presence in the virtually created environments is similar and sometimes even equal to the one in the real world. When looking at the most influential virtual reality headsets, based on sales statistics [8], and their main purpose, it becomes clear that virtual reality nowadays is primarily being used for gaming. Since this combination of virtual reality devices and gaming, which experiences a growing significance from a commercial and user-oriented point of view, has not been given much attention from a research perspective yet, the impact of virtual reality gaming on the user will be subject of this paper. More specifically, this paper evaluates whether virtual reality gaming evokes stronger brain activities than traditional gaming and how significant these are.

The paper is structured as follows. Based on the background, the problem statement and the resulting motivation of this paper will be illustrated in detail. The subsequent chapters will describe the experiment's methodology, its procedure and the collected data with a following analysis and interpretation. Furthermore, the author will use the collected results and their respective interpretations to suggest future research.

## Problem Statement

Examining the earlier mentioned sales statistics for games as well as for virtual reality devices, which are primarily used for gaming, one can easily identify an ongoing trend towards virtual reality gaming, which reaches an increasing number of people. Despite this trend and the fact that virtual reality gaming is no longer a niche market, there is only little research that has evaluated virtual reality gaming and its impacts on the user. Due to the growing circle of virtual reality gamers, understanding which impacts, especially on the neuronal processes the user has to expect, is of utmost importance.

Even though virtual reality has a long history from a technical perspective, its utilization as a mass-market-oriented gaming device has to be seen as a novum and should therefore, just like any other new technology that includes human-computer-interactions, be rigorously tested. Due to the lack of research findings for virtual reality gaming and its impacts on the user, the gaming industry assumes that findings from the related field of (traditional) videogame research are generally valid and can be transferred. As a case in point, the classification of videogames can be named. As mentioned earlier, traditional videogames have been rigorously tested and various impacts on the user have been identified. Therefore, videogames are classified in regards to their suitability for different age groups. The classification mainly focuses on the content of the videogame. Even though virtual reality gaming is a new medium, the same classification approaches are applied, which means it is based on the assumption that past findings can be applied without further virtual reality specific amendments. The fact that modern virtual reality devices, as a new technology for gaming, has not been rigorously tested yet and therefore defines its suitability for different target groups based on assumptions, leaves an increased risk for its users behind.

## Motivation

The existing knowledge gap which impact virtual reality gaming has on the user and to what extend this impact is different to traditional videogaming, is of great significance. While the impacts of traditional videogaming have been extensively tested, resulting in standards and classifications, there are no virtual reality specific mechanisms that are applied. No findings have been authored yet whether the past findings from the traditional videogaming area of research can be transferred and applied without further adjustments. Creating new knowledge and gaining an understanding of the differences between virtual reality and traditional video gaming, especially in regards to the neuronal activities of the gamer, will benefit various target groups. First at all, this knowledge will result in a virtual reality gaming specific classification of games, which allows a better definition of target- and risk groups and therefore mitigates the risk of negative side effects for the gamer. Furthermore, developers of games should be able to make use of a better understanding of neuronal distinctions of virtual reality gaming and incorporate these findings into their projects. Ultimately, they can reduce any features or contents in their games that are more likely to cause negatively associated neuronal activities and therefore reach a greater, unconscious acceptance of their games. The developers who are able to understand virtual reality gaming in the context of neuronal differences to traditional gaming will be more likely to publish successful games, because their games will then strive to create only positively associated reactions.

In conclusion the motivation for this paper is to understand possible differences between virtual reality gaming and traditional gaming and, based on the results, then reject the unfounded and generally applied assumption in the industry. Therefore, the long-term motivation of the present paper is, that an awareness of the differences and a better understanding of these will result in an increased quality of virtual reality games. By mitigating possible side effects, the overall acceptance of virtual reality as a new technology should increase accordingly.

## METHODOLOGY

### Introduction

The underlying work of this paper is a small pilot study that was conducted in the virtual reality laboratory at Xi'an Jiaotong Liverpool University in Suzhou, China. The pilot study aims to answer the research questions if virtual reality gaming evokes different brain activities than traditional gaming and how significant these differences are. During the pilot study the brain waves of the participants were recorded with a neuroheadset. Participants played on a tv-screen (referred to as traditional gaming) as well as with a Sony virtual reality headset. To enrich the obtained brain wave data, they were also asked to answer a questionnaire. The experiment will be illustrated in detail in the next chapters.

### Questionnaire Development

The aim of the additional questionnaire was to further understand the evoked brain activities from participants and to support the brain wave measurements. While the neuroheadset records electrical activities of the human brain and the software, with its underlying algorithm, converts these into a dashboard with six metrics, the questionnaire was designed to capture the conscious reactions of the participants. Finally, the results from the questionnaire will support the brain wave results.

The questionnaire is developed from findings of the study 'Measuring and Defining the Experience of Immersion in Games' [4]. As the title of their work states, their research goal was to find a way to measure immersion, a concept that is strongly related to the concept of presence, which was introduced earlier. Due to the different research objectives, the questionnaire developed within their study had to be altered to be consistent with the results obtained from the brain wave recording sessions. The final questionnaire contains eighteen questions, asking the participants how strongly they agree to statements about their experience of playing the game. A five-point answer scale was used, 1 being strongly disagree (wording in the questionnaire 'Not at all') and 5 being strongly agree (wording in the questionnaire 'Very much so').

The questions in the questionnaire were selected and structured to meet the six metrics from the neuroheadset and the software. Thus, part one of the questionnaire included three questions to ask for the conscious and subjective engagement with the game. Part two also included three questions, but asking for the perceived level of excitement while playing. The three questions in the third part were selected to ask the participant to which level he/she was interested in the game. Part four, with its three questions aimed to receive the level of perceived relaxation, while part five and six asked for the perceived levels of stress and focus.

### Equipment and Materials

Despite utilizing only a small-scaled pilot study for this paper, various materials had to be studied and applied. Since this paper's research question is whether virtual reality gaming evokes different brain activities than traditional gaming (on a tv-screen), the core materials of this experiment were a Sony PlayStation 4, the Sony virtual reality headset, a Dual-Shock 4 controller, the standard Playstation 4 wired headset, four games, here Battlefield 1 (Release October 2016), Battlezone (Release October 2016), Horizon Zero Dawn (Release March 2017) and RIGS: Mechanized Combat League (Release October 2016) and a 40-inch Xiaomi LED Screen TV. To measure the respective brain activities the neuroheadset Emotiv Epoc+ was used (see Figure 1).

Figure 1. Emotiv Epoc+ Neuroheadset [12].

The free accompanying software Emotiv Xavier Controlpanel was used to record live metrics from the device. Data recorded with the neuroheadset is firstly analyzed in the cloud, using the company's algorithm and then represented in a dashboard with six metrics, see example screenshot in Figure 2.
To analyze the results Microsoft Excel was used.

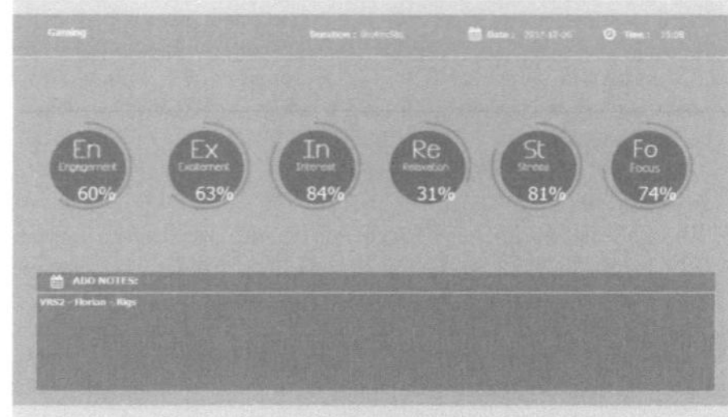


Figure 2. Controlpanel Result Dashboard (own Screenshot)

### Participants

Seven participants, six of them students at Xi'an Jiaotong Liverpool University in Suzhou, took part in this study. They had an average age of 25 (standard deviation = 2,16) and six of them were male and one participant was female. All of them had prior experience with PlayStation games but have never used virtual reality before. As the study is a pilot study, the author utilized the convenience sampling method. The participants didn't receive any payments or incentives for taking part in the experiment.

### Game Characteristics and Task

All four games were set up prior, thus participants were able to start playing immediately without dealing with menus or any other setting adjustments. Since all participants stated to have experience with games, the difficulty level was set to intermediate, if possible, and each participant was given a short verbal introduction to the game and its mechanics. The following will describe the situation for the participants in each game briefly.

Battlefield: The player finds himself playing a soldier in the first-person perspective, shooting at enemies.
Battelzone (Virtual Reality game): The player steers a tank in the first-person perspective, shooting at enemies.
Horizon Zero Dawn: The player plays as a warrior in an open world and shoots at enemies mainly with a bow.
Rigs (Virtual Reality game): The player steers a robot from the first-person perspective, again shooting at enemies

No specific task was given to the participants but all games were set in a state which required an immediate response from the player. The fact that all participants were familiar with videogames allowed them to jump into the action-packed situations immediately. In all four games the participant plays a virtual character that has to eliminate enemies by shooting at them. In all of the four in-game situations the participant will be confronted and attacked by enemies within the first minute and can only survive with a combination of weapon utilization and dodging enemy attacks. All of the four games were set up prior to the start of the experiment to ensure the highest possible level of similarity between them. These previously defined similar in-game situations as well as the limited playtime of five minutes reduces the significance and influence of factors like story of the game, character to be played or any other context-related parts. The main goal here was to create situations that, despite different games, feel highly similar for the participant.

### Procedure

To mitigate the risk of receiving noisy data from the neuroheadset a strict procedure was developed and adhered to. Participants were given an introduction to the experiment and after completing the consent form they were introduced to the Emotiv headset and the virtual reality headset. As a first step, the Emotiv headset was put on the participant's head. The Emotiv Xavier Controlpanel then indicates with a virtual scalp map whether the 16 saline based wet sensors are in the right position and if they can detect the brain's electrical activities. In order to start a recording session, the device has to be calibrated. This is to be done by defining the purpose of the recording, here "Gaming", followed by a short period of concentrating on a game, followed by a short period of just closed eyes. After this calibration, the participant started to play the first session.

The first session was a non-virtual-reality game session (NS1) playing the game Battlefiel 1. The recording within the software's controlpanel was started as well. After five minutes the recording and the gaming session were stopped and the participant had to answer the questionnaire for the first time, this time referring to the game that was just played. After answering the questionnaire, the second gaming and recording session were started. This session was a virtual-reality game session (VRS1) playing the game Battlezone. Again, after five minutes the sessions were stopped and the participant had to answer the questionnaire again, stating the perceived experiences during the game. This procedure was repeated once more resulting in a second non-virtual-reality game session (NS2, with Horizon Zero Dawn) and a second virtual reality session (VRS2, with Rigs), as well as two more questionnaires that were answered after each session. For each participant four questionnaires and four dashboards of recorded brain activities were collected.

### Methodological Problems

One of the problems that occurred at the very beginning of designing the experiment is based on the limited comparability of different games. While the first draft of this experiment intended to compare virtual reality with traditional gaming using the same game in both sessions, this could not be realized. As mentioned earlier, virtual reality gaming does represent a new medium, therefore the virtual reality games have to be specifically developed for it. For that reason, there are usually traditional videogames and virtual reality games on its own. Some traditional PlayStation 4 games though offer what is called 'VR-Compatible'. VR-Compatible often means the game supports the usage of the virtual reality headset, but does not have a full virtual reality support. The game 'Tekken 7' for example can be purchased as VR-Compatible version which then allows the player to play the game on a simulated large-scale display, creating the feeling of playing the game on a cinema-like screen. While this still uses the virtual reality headset including the head-tracking, it does not change the game mechanics from the traditional game, which makes it, what the author calls it, a hybrid game but not a virtual reality game. Thus, this limitation didn't allow a comparison that was based solely on one single game. The alternative experimental procedure therefore was designed to address this limitation. As mentioned earlier, the selected games, the prepared in-game situations and the play-time were adjusted for comparability reasons. Also, the author decided to increase the number of sessions from one to two per type (NS/VRS). By adjusting these factors, a high level of comparability within the experiment was created nonetheless.

Another problem that occurred was of physical nature since the Playstation virtual reality headset is not meant to be borne over another headset, in this case the Emotiv+ headset. While it is possible to wear both devices without decreasing the quality of the data received from the Emotiv+ headset, nor decreasing the quality of the perceived videogame image, it negatively impacts the overall comfort and can create a tense feeling around the head. This problem also contributed to the choice of only five minutes per play session.

## RESULTS

### Brainwave results

Seven participants, two non-virtual-reality and two virtual-reality sessions resulted in 28 dashboards (see Figure 2 for example), showing percentage scores for the six metrics engagement, excitement, interest, relaxation, stress and focus. The raw data consists of 168 percentage values (seven participants, 4 sessions each, recorded in six metrics).

For a better visualization of the data, boxplots were created for each session. Figure 3 shows this illustration. The data from each session, NS1, NS2, VRS1 and VRS2 is displayed as boxes and whiskers. While the bottom of each box represents quartile one and the top represents quartile three, the whiskers extending down represent the minimum value in each session and respectively the maximum value is represented by the whiskers extending up. The lines seperating the rectangles are the respective median values.

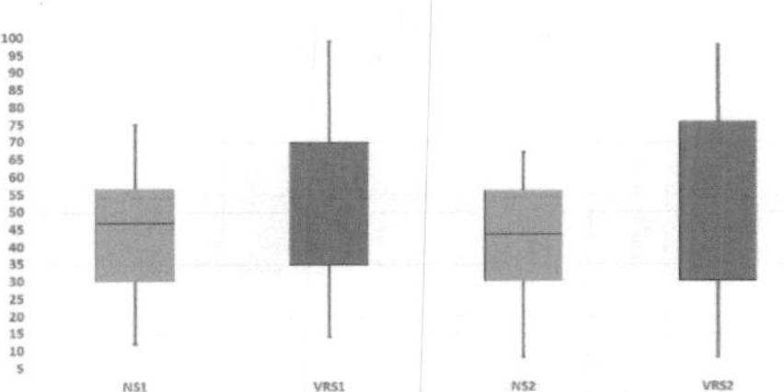

**Figure 3. Boxplot comparing NS1/VRS1 and NS2/VRS2 (own figure).**

[The first and the third one differ from the other two.]

Table 1 and Table 2 on the other hand show the mean results from NS1 (from all seven participants) compared with the results from VRS1 and the results from NS2 compared with VRS2 for each of the six metrics.

| Metric | NS2 | VRS2 |
|---|---|---|
| Engagement | 60,70 | 55,56 |
| Excitement | 28,70 | 42,00 |
| Interest | 61,00 | 76,86 |
| Relaxation | 31,56 | 31,70 |
| Stress | 49,56 | 73,29 |
| Focus | 35,13 | 53,00 |

**Table 1. Comparing mean values from NS1 and VRS1**

| Metric | NS1 | VRS1 |
|---|---|---|
| Engagement | 58,43 | 57,86 |
| Excitement | 23,29 | 34,70 |
| Interest | 59,56 | 82,56 |
| Relaxation | 30,43 | 31,43 |
| Stress | 43,43 | 77,13 |
| Focus | 31,28 | 50,13 |

**Table 2. Comparing mean values from NS2 and VRS2**

## Questionnaire Results

A similar approach was applied to the survey results. 18 questions, four questionnaires and seven participants resulted in 504 single values reflecting the subjective answers on the five-point scale. Table 3 shows the mean values for the 18 questions. Due to space constraints the two comparisons between NS1 and VRS1 values and between NS2 and VRS2 values were put into one table. Furthermore, the table does not contain the question wording but a reference to indicate which question evoked the particular mean values. The full questionnaire can be found in the appendix.

| Question | NS1 | VRS1 | NS2 | VRS2 |
|---|---|---|---|---|
| Question 1 | 3,28 | 4,13 | 3,00 | 4.70 |
| Question 2 | 3,28 | 4,00 | 3,14 | 4,43 |
| Question 3 | 3,00 | 3,56 | 2,43 | 4,13 |
| Question 4 | 3,00 | 4,13 | 2,70 | 4,28 |
| Question 5 | 3,13 | 4,28 | 2,70 | 4,28 |
| Question 6 | 3,00 | 3,70 | 2,56 | 4,28 |
| Question 7 | 2,70 | 4,00 | 2,56 | 3,86 |
| Question 8 | 2,13 | 3,70 | 2,43 | 4,13 |
| Question 9 | 2,00 | 4,00 | 2,13 | 3,86 |
| Question 10 | 4,43 | 1,56 | 3,56 | 1,43 |
| Question 11 | 3,00 | 3,56 | 3,13 | 3,56 |
| Question 12 | 3,00 | 3,29 | 2,70 | 3,00 |
| Question 13 | 1,29 | 2,86 | 2,00 | 4,00 |
| Question 14 | 1,43 | 4,13 | 1,56 | 4,56 |
| Question 15 | 1,43 | 4,28 | 2,13 | 4,56 |
| Question 16 | 2,43 | 3,86 | 2,13 | 3,86 |
| Question 17 | 3,00 | 4,70 | 2,70 | 4,28 |
| Question 18 | 2,56 | 3,56 | 2,70 | 3,86 |

**Table 3. Comparing mean values from NS1/2 and VRS1/2**

## DISCUSSION

### Analysis

While the research question of this paper is whether virtual reality gaming evokes stronger brain activities than traditional gaming, due to the six metrics from the Emotiv headset and the accordingly designed questionnaire, this can be broken down into six hypotheses.

H1: Virtual reality gaming evokes stronger engagement levels than traditional gaming.
H2: Virtual reality gaming evokes stronger excitement levels than traditional gaming.
H3: Virtual reality gaming evokes stronger interest levels than traditional gaming.
H4: Virtual reality gaming evokes weaker relaxation levels than traditional gaming.
H5: Virtual reality gaming evokes stronger stress levels than traditional gaming.
H6: Virtual reality gaming evokes stronger focus levels than traditional gaming.

To test these hypotheses, mean values for each of the six metrics were generated as a first step. This resulted in six tables, one for each metric, with two columns (NS (NS1+NS2) & VRS (VRS1+VRS2)) and seven rows with the mean values from the seven participants. While these mean values already give first indications, the paired t-Test is applied to all six tables for a more precise analysis. The paired t-Test can be performed within Excel with the built-in Data Analysis Tool. Table 4 shows the t-Test analysis for Hypothesis 5, based on the stress values from the Emotiv headset.

The usage of the paired t-Test in this case is suitable since the participants remain the same and one variable, here the type of gaming is being changed. Based on the research question and the six hypotheses, the H0-hypothesis for each of them is, that there is no significant difference between traditional and virtual reality gaming regarding measurable brain waves. Another way to express this is that the change of the independent variable does not cause a significant change of the dependent variable [10]. Table 4 shows various information that will be illustrated in further detail.

| t-Test: Paired Two Sample for Means | NS | VRS |
|---|---|---|
| Mean | 46,5 | 75,21429 |
| Variance | 51,66667 | 243,7381 |
| Observations | 7 | 7 |
| Pearson Correlation | 0,512762 | |
| Hypothesized Mean Difference | 0 | |
| df | 6 | |
| t Stat | -5,65748 | |
| $P(T_{\text{i}}=t)$ one-tail | 0,000655 | |
| t Critical one-tail | 1,94318 | |
| $P(T_{\text{i}}=t)$ two-tail | 0,00131 | |
| t Critical two-tail | 2,446912 | |

**Table 4. t-Test based on stress-values from NS1/2 and VRS1/2.**

The first two values in Table 4 show that the mean value for the virtual reality sessions is higher, which gives a first indication that the H0 hypothesis can be rejected and the alternative hypothesis, that a significant change exists, is supportable. The mean Pearson Correlation coefficient of 0,513 represents a moderately strong correlation, when applying the widely accepted rule of thumb for the Pearson coefficient. To further support Hypothesis 5, that virtual reality gaming evokes stronger stress levels than traditional gaming, the absolute value of the t Stat result is compared to the t Critical one-tail value. The value of -5,657484434 shows again that the second mean value is higher than the first one. In a next step it is tested whether this is caused by a standard error or if we can actually confirm a significant difference. Since the absolute value (here 5,657484434) is higher than the t Critical one-tail value (here 1,943180281) we can reject the H0 hypothesis. In a last analysis step the value in the row 'P($T_i$=t) one-tail' is compared with the previously set Alpha of 0,05. This value is significantly lower than 0,05, which ultimately shows that the H0 hypothesis can be rejected and Hypothesis 5 can be confirmed. The results of the paired t-Test for the stress values show that virtual reality gaming evokes stronger stress levels.

This can be confirmed by the same approach applied on the questionnaire results. Again, the relevant results from the t-Test show that virtual reality gaming evokes stronger stress levels than traditional gaming, this time based on the conscious perception of the participants. Table 5 shows the questionnaire equivalent of Table 4.

| t-Test: Paired Two Sample for Means | NS | VRS |
|---|---|---|
| Mean | 1,642857 | 4,071429 |
| Variance | 0,281746 | 0,239418 |
| Observations | 7 | 7 |
| Pearson Correlation | -0,08149 | |
| Hypothesized Mean Difference | 0 | |
| df | 6 | |
| t Stat | -8,55965 | |
| P($T_i$=t) one-tail | 6,98E-05 | |
| t Critical one-tail | 1,94318 | |
| P($T_i$=t) two-tail | 0,00014 | |
| t Critical two-tail | 2,446912 | |

**Table 5. t-Test based on stress questions from NS1/2 and VRS1/2.**

In total the analysis consists of 12 t-Test tables comparing the seven participant's results. Six for the obtained brain wave data, one for each metric and six for the respective questions linked to these six metrics. Therefore, the remaining five hypotheses can be evaluated whether they can be confirmed or rejected, by also applying the t-Test to the brain wave results and the questionnaire results. The results of these analyses are as follows.

Hypothesis 1, Engagement:
Emotiv results show that Hypothesis 1 cannot be confirmed. Questionnaire results show that Hypothesis 1 can be confirmed.
Hypothesis 2, Excitement:
Emotiv results show that Hypothesis 2 can be confirmed. Questionnaire results also show that Hypothesis 2 can be confirmed.

Hypothesis 3, Interest:
Emotiv results show that Hypothesis 3 can be confirmed. Questionnaire results also show that Hypothesis 3 can be confirmed.
Hypothesis 4, Relaxation:
Emotiv results show that Hypothesis 4 cannot be confirmed. Questionnaire results show that Hypothesis 4 can be confirmed.
Hypothesis 6, Focus:
Emotiv results show that Hypothesis 6 can be confirmed. Questionnaire results also show that Hypothesis 6 can be confirmed.

**Interpretation of results**
The results and analyses from the preceding chapters show that virtual reality gaming does evoke different brain activities and that they are of some significance. The first indications which were made visible in Figure 3 as well as Table 1 and Table 2, were confirmed with the in-depth t-Test analyses. Hypothesis 2, 3, 5 and 6 were confirmed consistently, according to the conscious and unconscious parts of the experiment. Hypothesis 1 and Hypothesis 4 generated inconsistent results from the two experimental parts. Overall in four out of six cases the Emotiv brain wave results and the questionnaire results consistently supported either a confirmation or rejection of the hypothesis.
Hypothesis 1 couldn't be consistently confirmed nor rejected. While the Emotiv neuroheadset results show that there is no stronger level of engagement, the majority of participants (6 out of 7) rated the conscious engagement level as stronger for the virtual reality sessions. Hypothesis 4 couldn't be confirmed nor rejected either due to inconsistent results. While the t-Test applied to the Emotiv results indicates no significant and systematic differences, the questionnaire results show very little difference.

Besides that, there has been enough evidence to confirm that virtual reality evokes stronger levels of excitement, interest, stress and focus. The most significant difference was identified for the stress level, see Figure 4. The mean value for the virtual reality sessions shows a difference of +61,75% for the brain wave analysis and +147,83% for the questionnaire analysis, compared to the normal gaming sessions.

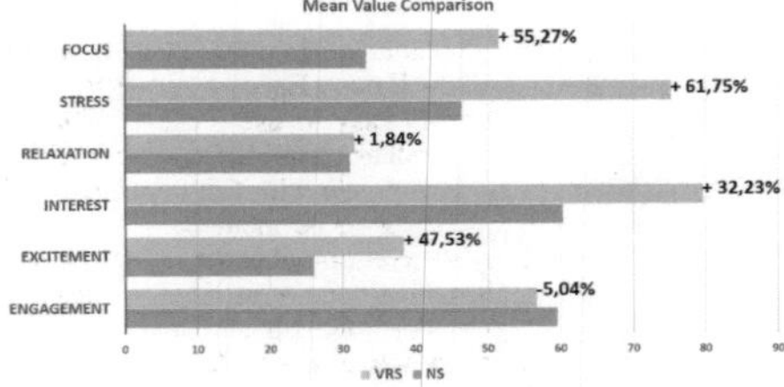


**Figure 4. Mean value metrics comparison (own figure).**

[The upper chart stands for VRS, the lower one for NS.]

That shows, that even though the situations within the games were selected to be as similar as possible and despite the lack of a task to archive, being confronted with enemies that attack the player creates a significantly greater stress level in virtual reality. This correlates with the findings of other non-gaming related virtual reality studies, which showed, that virtual reality environments have the potential to be perceived nearly identical to the real-life situations and therefore cause stronger reactions. Based on the results of this paper it can be said that switching from a tv screen to a virtual reality device, causes the players to experience a stronger feeling of presence and therefore show stronger reactions, consciously and unconsciously. That said, the mainly higher mean results show that virtual reality has a greater potential to blur the distinctions between virtual and real environments for the user, making the user experience real-life-like reactions to the virtual situations. While this sounds like the dream of most gamers, to find themselves in real-life-like game environments, it does come along with rather negatively associated aspects like the increased level of stress while playing. The statements from three participants of this study who mentioned after the experiment that virtual reality gaming was an interesting but exhausting experience for them, supports that interpretation.

At this point it has to be mentioned again that physical problems occurred during the experiment, when fitting the virtual reality headset over the neuroheadset. Three participants actually stated to experience a tense feeling around the head after being equipped with both headsets. This perceived tense feeling might have at least influenced the higher stress results in this experiment. Furthermore, two participants felt a strong dizziness after the virtual reality sessions and one even expressed that he might have to throw up. All participants showed at least some signs of relief when being allowed to take off the Emotiv neuroheadset again.

In summary, the research question whether virtual reality evokes stronger brain activities was confirmed and beyond that was illustrated in more detail, using six different metrics and evaluating six hypotheses.

## Critical Reflection and Limitations

Even though the small scaled pilot study did generate results that allowed answering the research question and confirming the initial hypothesis, the scientific scope of this paper has to be reviewed critically due to various limitations.

The majority of limitations arise from the equipment used. First at all, even though the Emotiv Epoc+ became quite popular as an inexpensive and easy to use brain-computer interface, in the end it is a rather simplified consumer market neuro-device that cannot compete with fully equipped EEG laboratories. Secondly, it has to be mentioned that for this study the full potential of the Emotiv headset wasn't utilized. While the headset, by standard, captures EEG raw data, obtaining it from the headset and being able to analyze it is tied to an Emotiv subscription that comes with a continuous fee. Beyond that, obtaining and analyzing raw EEG data requires profound knowledge in this field of research. Due to time constraints for this paper, the simplified approach, here the Emotiv Xavier Controlpanel, was therefore used. While the results from this simplified approach were successfully used to confirm the initial research question, one has to bear in mind possible lower accuracies.

Another downfall of this limited approach is a lack of transparency. After finishing the recording session with the neuroheadset, the collected data is transmitted to the Emotiv cloud servers for the analysis. The underlying algorithms and logics are not publicly visible and can't be reviewed for its suitability for the given scenario. Thus, the simplified approach of receiving easily analyzable data can be seen as a major advantage as well as a major disadvantage, impacting the general validity of this paper. Furthermore, it has to be mentioned again that despite all the efforts to place the participants in highly similar in-game situations to increase the comparability, having different games still carries the risk of subjective game-preferences influencing the results.

Overall, for a successful full-scale study the raw EEG data should be object of the analyses rather than the dashboards, the selected games should vary as little as possible and the number of participants should be higher than in the present pilot study.

## Implications and Research Agenda

The findings of this paper indicate that, despite the long history of virtual-reality from a research and technical perspective, it should be handled with the rigorous testing that is usually applied to new technologies. In the course of this paper it became clear that virtual reality as a mass market device and application still leaves important questions unanswered. As this paper functions as groundwork, it delivered first results to point not only researchers but also producers in the virtual-reality industry towards these questions.

This paper confirmed that virtual reality evokes stronger brain activities that, from a gamer's point of view, can be categorized as positive, like a greater level of perceived excitement, or that can be categorized as negative, like a greater level of stress. Thus, further research can analyze how to maintain or even boost the positive effects and how to eliminate possible negative effects of virtual reality gaming. Developers might further examine which parts of their game evoke the various neuronal responses to focus on the positive ones only.

Furthermore, it also became clear that the past gaming studies and their respective results can't just be extrapolated for virtual reality gaming since the differences are too significant. Further studies have to show what the impact of the stronger brain activities will be on regular gamers or on gamers that already showed certain signs of gaming addiction. Will the higher level of excitement result in a greater risk of drifting into a game addiction or will this be balanced out by the higher level of stress? What long-term effects on regular virtual reality gamers can be identified and should the industry establish a distinguished classification framework for virtual reality games? These are possible research questions for further studies. Pointing at producers and publishers of virtual reality games on the other hand, questions have to be raised whether the stronger neuronal reactions should be made a subject of discussion to create greater awareness among gamers.

## CONCLUSION

In this paper the existing knowledge gap for virtual reality gaming has been discussed, from a scientific perspective as well as from a practical and rather user-oriented perspective. How significant the differences between virtual reality gaming and traditional gaming can be has been investigated with a small-scale pilot study. Virtual reality does not only create a stronger reaction that is based on conscious feelings, but also the associated reactions in the user's brain were found to be mainly stronger. The questionnaire developed was predominantly successful in underlining the obtained neuronal results and supported consistent conclusions for the six hypotheses.

The second and main part of the experiment, using a neuro-headset to measure the brain activities of participants during gaming sessions, was also successfully applied, even though two of the derived hypotheses couldn't be confirmed. The author's own feeling of being left with a rather exhausted and unpleasant feeling after playing virtual reality games was analyzed and the results were prepared in comparing tables. The results of this paper help to create a well-founded awareness on what users can and should expect from modern virtual reality gaming. One aim of this paper was to understand whether past findings for traditional games can be transferred or extrapolated to the field of virtual reality gaming. The author of this paper claims that the identified differences are too significant to stick to the outdated assumptions. Virtual reality gaming therefore should be rigorously tested and existing classification mechanisms should be adjusted accordingly. The fact that various participants of the small pilot study expressed an uneasy feeling afterwards just amplifies the findings and supports the author's conclusions.

Overall, it is evident that the previous virtual reality studies and the present study harmonize, showing that virtual reality has a potential to evoke user reactions that are comparably to real-life reactions. Even though that has always been the goal of games, to allow the user to experience situations as realistically as possible or to experience situations impossible in the real world, negative emotions like stress and uneasy feelings also run high when using virtual reality. However, as with most new technologies developers are still adopting it and sounding out possibilities. Thus, this paper aims to help during this process by creating the necessary awareness and guiding to further research.


## REFERENCES

1. Biocca, F. The cyborgs dilemma: Embodiment in virtual environments. *Proc. of the 2nd International Conference on Cognitive Technology* (1997), 12–26.

2. User Experience und Experience Design - Konzepte und Herausforderungen.
`https://pdfs.semanticscholar.org/22b4/`
`0bf10000e9aa9f323c1a62c6792ce70ed4ba.pdf`.

3. Gaming for Health: Systematic Review and Meta-analysis of the Physical and Cognitive Effects of Active Computer Gaming in Older Adults.
`https://academic.oup.com/ptj/article-abstract/`
`doi/10.1093/ptj/pzx088/4097725?redirectedFrom=`
`fulltext`.

4. Measuring and defining the experience of immersion in games. `http://citeseerx.ist.psu.edu/viewdoc/`
`download?doi=10.1.1.157.4129&rep=rep1&type=pdf`.

5. Jeong, E. J., Kim, D. J., Lee, D. M., and Lee, H. R. A study of digital game addiction from aggression, loneliness and depression perspectives. *System Sciences (HICSS)* (2016), 3769–3780.

6. Mandryk, R., Birk, M. V., Lobel, A., van Rooij, M., Granic, I., and Abeele, V. V. Games for the assessment and treatment of mental health. *Proc. of CHI PLAY '17 Extended Abstracts* (2017), 673–678.

7. Prognose zum Umsatz im Markt fuer Videogames weltweit von 2011 bis 2020 nach Segment (in Milliarden US-Dollar). `https://de.statista.com/`
`statistik/daten/studie/160518/umfrage/`
`prognostizierter-umsatz-in-der-weltweiten-`
`videogames-branche/`.

8. Virtual reality device unit shipments worldwide by vendor/brand in 2016 (in 1,000s)).
`https://www.statista.com/statistics/705476/`
`global-virtual-reality-device-shipments-by-vendor/`.

9. Virtual reality software and hardware market size worldwide from 2016 to 2020 (in billion U.S. dollars). `https://www.statista.com/statistics/528779/`
`virtual-reality-market-size-worldwide/`.

10. Student. The probable error of a mean. *Biometrika 6*, 1 (1908), 1–25.

11. Measuring (Tele)Presence: The Temple Presence Inventory. `http://citeseerx.ist.psu.edu/viewdoc/`
`download?doi=10.1.1.231.4747&rep=rep1&type=pdf`.

12. Emotiv Epoc+ Official Website https://www.emotiv.com/epoc/

# Virtual reality gaming evokes stronger brain activities than traditional gaming.

**Marvin Brucker**
Xi'An Jiaotong-Liverpool University
Suzhou

## ABSTRACT

Virtual reality devices are reaching into the consumer mass market and are on the edge of influencing more and more areas of our daily life. One of the most common applications for virtual reality today can be found in the gaming industry. Due to the novel character of these modern virtual reality devices and the potential of reaching more people than all other head-mounted devices before, the effects of virtual reality gaming on the human brain is of great interest. Thus, this paper evaluates whether virtual reality gaming evokes different brain activities than traditional gaming and how significant these differences are.

## Author Keywords
Virtual reality; gaming; brain activities.

## ACM Classification Keywords
I.3.7.[Computer Graphics]: Three-Dimensional Graphics and Realism - Virtual Reality; H.5.2 [Information interfaces and presentation]: User Interfaces - Graphical user interfaces.

## INTRODUCTION
Virtual worlds and artificially created images have an increasing impact on our daily lives and one can undoubtedly say that virtual reality is a future technology which will have an increasing impact on various application areas like research, industry and consumer electronics. Due to the rapid development of display hardware, hardware for interactions and tracking systems, todays virtual reality applications, which a few years ago were only available within specialized research labs, are now widely available to researchers and consumers. Therefore, virtual reality today can be found in various application fields such as advertisement, movies, computer games and professional applications that enable developers to visualize, evaluate and refine their prototypes in real-time.

Due to the novel character of virtual reality as a scientific field, which is highly dependent on the development of computing hardware and the utilized methods, a uniform definition of what virtual reality actually is has not been authored yet. Even though to say "if you move your head and nothing happens it ain't VR" [5] is just a very limited approach of defining virtual reality, it does show a general tendency of most of the definitions, which erroneously and predominantly refer to the technological aspects of virtual reality for defining it. Having said this, if a definition focuses mainly on technological components, it might inherit a decreasing applicability over time due to the hard- and software advancements.

A rather future-proof definition therefore is "Virtual Reality refers to immersive, interactive, multi-sensory, viewer-centered, three-dimensional computer-generated environments and the combination of technologies required to build these environments" [6]. Ultimately with virtual reality, humans are in the center of a virtually created and defined world and the resulting mixed-reality, consisting of reality and virtuality [18] constantly produces new applications that will emerge into the mass market and allow the user to experience virtual interactions that appear to be as authentic as possible. One of the main areas of virtual reality implementations is the field of digital gaming, where this new level of authenticity meets a high level of interactivity with the virtual world. Whether and how strongly this combination of gaming and virtual reality impacts the user's neuronal processes will be the subject of this paper by measuring the brain activities.

## HISTORY OF VIRTUAL REALITY
The fundamentals for creating these virtual interaction experiences exist for several decades. Again, there is no unity among researchers about what can be seen as the beginning of the virtual reality history but most refer to Ivan Sutherland who, in the course of his immersive technologies research activities, wrote the paper "The Ultimate Display" in 1965. This paper can be seen as a first step of combining the computer with the design, the construction, the navigation and the experience of virtual worlds [36]. His invention of the "Head-Mounted Display", a data helmet, enabled the viewer to experience a simulated simple 3D-environment [1].

The next big step in virtual reality can be found two decades later. Thomas Zimmermann and Jaron Lanier did not only establish the company "Visual Programming Languages", which was one of the first to penetrate the consumer market with virtual reality products, but Jaron Lanier was also the first to actually use the term virtual reality. They developed the "DataGlove", a data glove that was equipped with fiber optic on the back of the hand, to capture finger data and they also developed the data helmet "EyePhone", which can be seen as a clear continuation of Sutherlands head-mounted display with the aim to create an immersive virtual reality experience [35].

After some financially unsuccessful attempts to introduce virtual reality to the gaming section, e.g. Sega VR in 1991 [28] and Nintendo's "Virtual Boy" in 1995 [22], a big step forward was made in 2001 with the SAS3TM. Due to advancements in PC hardware, the "SAS cube" was the first device actually coming close to the earlier mentioned "ultimate display" and allowed the creation of what can be seen as a virtual landscape [26].

Following Sutherlands work to create the ultimate display, most of the developed devices were mount-headed displays (or video glasses), thus visual output devices, which are borne in front of the eyes and, as long as no form of augmented reality extension is included, do not allow any penetration of light or pictures from the real world, in order not to interfere with the virtually generated world. This way, the necessary two images are not provided outside of the device and then assigned to the correct eye within the device, but two smaller displays are used in the device to create two stereoscopic pictures that are directly fed to the correct eye [7].

The principal of stereoscopy makes use of the ability of the human perception system to process separate pictures, that are presented to the left and the right eye, but differ slightly on the viewpoint. This way media can create binocular depth information and therefore deceive the visual organ with virtual objects and distances. The so called stereoscopic head-mounted displays in combination with head tracking have been the primary choice for the visualization of virtual environments during the virtual reality research in the 1990s [3] and still are.

While most of the research has been done in the 1990s, most of the devices were primarily reserved for research purposes and did not find their way to the mass market. The breakthrough of natural input devices in general, with speech- and gesture recognition for consumer electronics took place as late as 2005 with smart TVs and video game consoles like the Nintendo Wii or Microsoft Xbox 360 Kinect.

This initiated trend towards more natural interactions, the already existing research in the field of stereoscopic head-mounted displays as well as the increased computing power paved the way for a new era of virtual reality devices and applications within the consumer market. Even though the choice for the visualization remains the same, the differences are tremendous. The quality of the shown images has improved significantly and the level of immersion reachable today cannot be simply described anymore and can only be fully understood by actually wearing one of the modern virtual reality devices [39].

This trend can be confirmed by evaluating recent quantitative research regarding the virtual reality market. In 2016, Occulus Rift and the Sony Playstation Virtual Reality were released and approximately seven million virtual reality headsets were sold. According to estimations, by 2020 37 million headsets will be sold [32]. The most important devices, also based on sales figures from 2016, are Samsung Gear, Playstation Virtual Reality, Google Daydream and Oculus Rift [33]. The virtual reality market as a whole, so revenue from virtual reality soft- and hardware summed up, shows an estimated increase from 3,7 billion US-Dollar in 2016 to 40,4 billion US-Dollar by 2020 [34].

These figures also underpin that, even though virtual reality does have a long history, the medium itself as a mass market suitable one, is a relatively new field of application. Not only isn't there a uniform definition, but also the scientific exploration of these rather modern virtual reality devices and applications and their impact on the user is still in the early stages.

## INTRODUCTION TO GAMING RESEARCH

Since the 1970s videogames, regardless of their platform, have become a widely distributed and relevant media phenomenon within most modern societies, impacting the daily lives of children and adolescents as well as adults. Due to their unique level of interactivity among other forms of media consumption, videogames have changed and expanded the way we experience and explore medial presentations.

Despite the growing market and growing audience for videogames [31] ande despite a high social acceptance among young users, the overall public response remains ambiguous [10]. This ambivalence can be easily recognized when reviewing the broad research within the field of gaming. Studies that are related to gaming in any way, engage in countless ways with the medium games but one frequent phenomenon is that due to the short half-life of the popularity of many released games, researchers are driven to make reductions or abstractions to make reasonably and generally valid statements.

While the fields of psychology and media education [11] have a particular focus on problem-oriented aspects, such as addictive- and violence-fostering effects [16] and media science rather focuses on the picture and narrative composition of the game [27], social sciences so far have been mainly interested in usage patterns of gamers [8]. The results of the studies arising from different fields are as ambivalent as the public opinion.

When searching for gaming studies, one always stumbles across the great amount of studies that either underpin particular negative effects of gaming, just as gaming addiction [16, 9, 2] or that show, based on neuronal examinations, how games that involve violence can negatively impact the gamer's behavior patterns [17]. Furthermore there are numerous studies that, in their core, try to answer the question whether digital gaming is healthy and if, to what extent. Again, when reviewing studies from this particular area one will inevitable come across studies that show the (little) positive impacts [14, 20] as well as the negative impacts of gaming [15].

These kind of studies, that attribute certain health benefits or issues to gaming are widely picked up by the media, which then also utilizes the previously mentioned methods of reduction and abstraction to deliver generally valid statements [12, 4, 38, 25, 24].

It is noticeable that the area of gaming research is diverse, contains of nearly an innumerable amount of published studies and is publicly discussed due to an occasional extensive media coverage. Since the medium of virtual reality can be seen as a next evolutionary step in gaming and in media consumption in general, due to the increased level of interaction, the existing research work for virtual reality will be reviewed as well.

## INTRODUCTION TO VIRTUAL REALITY RESEARCH

Compared to digital gaming in general, the widespread of virtual reality devices and applications in the consumer market can be seen as a novelty. The commercially significant prevalence of virtual reality is only a few years old and therefore it is not surprisingly that papers examining the impacts of virtual reality on the user or in this case the gamer, are rare and this area of research is in an early stage (if one differentiates to the technology-oriented research of the past decades). But studies can be found that measure, in one or another way, physiological processes to conclude the impacts of virtual reality systems on the cognitive and emotional conditions of the user and to deliver quantitative indications to what extent this user experiences a presence within the virtual environment.

Meehan for example was able to find evidence that the heart rate and the skin conductance of subjects increased significantly when exposed to a virtual pit within a virtual environment [21]. Similar results were obtained by Simeonov when psychological and physiological changes shown by subjects were measured. Again, the scenario was related to height and a risk of falling and the subjects were exposed to a real-life situation and a virtual reality situation. Both situations were designed to be similar and measures were taken in both of them for further analysis and comparison. The subjects showed, based on the measured indicators, similar levels of anxiety and risk evaluation, despite a clear understanding that one of the situations was delivered completely by technological means [30].

Studies that compare real life situations with similar situations in a virtual environment are an emerging field of research, especially since recent results show the possible therapeutic effects of it [40]. Pertaub showed that even phobias that are not related to physical fears like the fear of falling, but rater social fears like the fear of public speaking, simulated in a virtual environment, were perceived as nearly identical as the real life situation by subjects. It is important to note that this reflects the power of virtual reality since it evokes a response, a response of presence, that is significantly similar to that of reality [23].

Usually the main effort of these studies are to compare reality with virtual reality with the goal to examine and treat people that suffer from phobias or even post-traumatic stress [19]. Additional studies were able to show that virtual reality environments can help to stimulate and treat injured brain areas and therefore verifiably increase the performance of some brain areas [29].

Further studies can be found that examine EEG data from subjects that are interacting with virtual reality environments to determine which factors impact the users perception in order to create a stronger feeling of presence. For further experiments that evaluate virtual reality environments and user's perception, the results from Havranek can help determining the experiment framework. He was able to show that both, a first-person perspective as well as experiencing the virtual environment in an interactive manner results in EEG signals that can be translated into a stronger presence feeling, compared to a third-person perspective and a passive spectator-like experience [13].

All these studies show that virtual reality has become an interesting field for researchers of various origins, especially with a medical background. These new overlapping research fields already created many possible applications but it is noticeable that most of these applications, studies and the associated results are very specific, in other words, specific environments are created to reproduce this same specific situation from the real word to then deliver it to the subject, either for collecting data for further analysis or to treat existing diseases.

The biggest advantage of virtual reality from a researchers perspective is the fact that it allows to create a safe research environment and therefore confront subjects with situations that were not possible before. But while examining these studies it is noticeable that the results only support one specific purpose and usually do not allow any generally valid statements about the impact of virtual reality on the users. A research team from UCLA came to the conclusion that this research area still demands further research, when they found that "the pattern of activity in a brain region involved in spatial learning in the virtual world is completely different than when it processes activity in the real world" [37].

## PROBLEM STATEMENT

In the preceding chapters it has become clear that there are numerous research results for the digital gaming area. Results from psychological, media educational, media scientific or sociological points of view, results with an image-composition-focused or narrative scientific background. The same applies to the rather young field of virtual reality research. Next to research about the required or supportive conditions that enable and strengthen the presence feeling, research exist that brings virtual reality and medical treatments together by examining the impact of virtual reality applications as treatment for brain injuries and mental diseases. Strikingly many similar research exist for gaming in general and virtual reality in general, studies that examine neuronal differences between traditional digital gaming and virtual reality gaming though are rare to nonexistent.

When examining the mentioned studies for the traditional digital gaming it is noticeable that it is attributed with strong impacts on the user, both positive as well as negative impacts, especially on regular gamers. This observation is even stronger when only paying attention to the studies that show one or another negative impact of regular digital gaming. The bottom line here is, digital gaming can have tremendous impacts on users.

Now reviewing the virtual reality research, the bottom line is often the creation of a virtual environment that shows the unprecedented power of virtual reality, a power that is strong enough to have tremendous impacts in therapeutics and neuroscience.

Conclusively the problem right now can be seen as a gap between the two research areas, leaving the important questions, whether virtual reality gaming evokes different brain activities than traditional gaming and how significant these differences are, unanswered. Neither consumers nor researchers at this point know what to expect when adding up the possible negative effects of traditional digital gaming, such as addiction potentials, and the power of virtual reality applications to reach new levels of user presence and user influence. Especially due to the shift from virtual reality as a niche product to a mass market product, these questions shouldnt be unanswered.

## MOTIVATION

The mentioned problem of an existing gap between the two research areas leads to questions that have to be raised and evaluated, such as: what will be the impact on users when bringing together virtual reality and gaming, especially on regular users? Are the results from digital gaming research that were introduced earlier still valid when virtual reality devices are applied and if so are the results just amplified or can we expect completely different results due to completely different brain activities when interacting with virtual reality environments?

Especially due to the numerous research that were able to show the risks of gaming addiction or research that showed how violent games can negatively impact the gamer's behaviors, answering the questions what impact virtual reality on gamers can have is of significant relevance.

As this paper aims to establish a new area of research, that combines the already existing areas of gaming research and virtual reality research, the first important part of this paper is to create an awareness of the existing gap between both areas. Consequentially this paper underpins the need for establishing this research area, by raising questions that have not been raised nor answered yet. This paper's limitation is that it is not able to answer all of the raised questions. Despite this limitation that the results can only be utilized to answer some of these questions, the author identifies the paper's potential to serve as a starting point for further research as its core strength. Therefore, the motivation of this paper is to create awareness that a lot of research has to be done, raise important questions, function as groundwork and deliver results that not only point at further descriptive research but also raise even more questions that still have to be rigorously tested.

## METHODOLOGY

In this paper the neuroheadset Emotiv Epoc+ will be used to measure the brain activities. The free accompanying software Emotiv Xavier Controlpanel allows to see and record six different live metrics from the device. User engagement, excitement, interest, relaxation, stress and focus. These metrics are monitored live by the software and will be recorded for different subjects. The result of each recording is a dashboard showing the strengths of the six metrics in percentage numbers. As the study is a pilot study, the author utilizes the convenience sampling method. Subjects will play the same game on a Sony Playstation 4, once by using the virtual reality device and once without the device but both times wearing the Emotiv neuroheadset and being connected to the recording software. To enrich the quantitative findings, the subjects will additionally take part in a post-experiment survey.

This methodology, especially the choice of the neuroheadset and the sampling method, are suitable for a pilot study, due to the relatively easy handling and the less expenditure. Furthermore, the chosen methodology meets the motivation of this paper, to function as descriptive research groundwork rather than answering all related questions.

That said, the scientific scope of this paper has to be reviewed critically for the exact same reason. These limitations mainly arise from the choice of the neuroheadset and the choice of software for the subsequent analysis. While the Emotiv Epoc+ became quite popular among mainly hobby-researchers and early adopters for being an inexpensive and easy to handle brain-computer interface, researchers that focus on the acquisition and analysis of EEG data usually use fully equipped laboratories rather than anything close to these consumer market neuro-devices. Secondly, even though the Emotiv neuroheadset is capable of recording raw EEG data, obtaining the data from the headset and analyzing it afterwards is not only tied to an Emotiv subscription, but moreover requires profound knowledge and research efforts in the areas of obtaining and filtering EEG data, identifying common EEG data patterns and interpreting raw findings. The chosen free software Emotiv Xavier Controlpanel on the other hand does not require most of these tasks and enables a simple data acquisition and interpretation. While this choice was the best considering the limited time frame for this project, its results and conclusions have to be reviewed critically due to the limited scope arising from possible lower accuracies.

## EXPECTED OUTCOMES AND CONTRIBUTIONS

It is expected that both results, the one from the neuroheadset recordings and the results from the survey will show a different reaction to virtual reality gaming, conscious- and unconsciously. Additionally, it is expected that the brain activities are in general on a stronger level and the average of subjectively felt immersion among the subjects is expected to be stronger as well. The results are expected to be strong and consistent enough to point towards further research that, by eliminating the limitations of this paper's methodology, has the potential to examine if stronger brain activities evoked by virtual reality gaming conclusively leads to further effects like a stronger perception of exhaustion for example or if a boosting effect of general gaming impacts can be found.

While this paper aims to not only be a groundwork in this field but also aims to answer some of the raised questions, the above mentioned outcomes are only expected outcomes at this point. As stated in the previous chapters, there have not been any research bringing these two areas together so the outcomes may point at complete different finding or even no findings at all. Neurophysicists at the UCLA were surprised to which degree their results from a virtual reality experiment diverged from what they had expected. The fact, that this pilot study utilizes a rather high-level software to analyze the obtained data, which will finally result in a dashboard-like result, will mitigate the risk of ending up with no findings at all.

## REFERENCES


1. Adams, R., and Merklinghaus, D. P. Augmenting virtual reality. *Military Technology 38*, 12 (2014), 16–24.

2. Batthyny, D., Mueller, K., Benker, F., and Woelfling, K. Computerspielverhalten: Klinische merkmale von abhaengigkeit und missbrauch bei jugendlichen. *Wiener klinische Wochenschrift 121*, 15 (2008), 502–509.

3. Biocca, F., and Delaney, B. Immersive virtual reality technology. *Communication in the age of virtual reality* (1995), 57–124.

4. Wie Computer das Gehirn beeinflussen. `http://www.zeit.de/online/2009/17/reiz-zu-reiz`.

5. Bryson, S. Virtual reality: A definition history - a personal essay. *NASA Ames Research Center* (1998/2017).

6. Cruz-Neira, C. Virtual reality overview. *SIGGRAPH 93 Course Notes 23* (1993), 1–18.

7. Cruz-Neira, C., Grimm, P., Herold, R., and Reiners, D. *VR-Ausgabegeraete*. Springer Berlin Heidelberg, 2013.

8. Festl, R., Scharkow, M., and Quandt, T. Digitales spielen als mediale unterhaltung. eine reprsentativstudie zur nutzung von computer- und videospielen in deutschland. *Media Perspektiven* (2010), 515–522.

9. Fritz, J., and Witting, T. Suche, sog, sucht: Was online-gaming problematisch machen kann. *Rausch ohne Drogen* (2009), 309–323.

10. Fromme, J., and Koenitz, C. Bildungspotenziale von computerspielen - ueberlegungen zur analyse und bildungstheoretischen einschaetzung eines hybriden medienphaenomens. *Perspektiven der Medienbildung* (2014), 235–286.

11. Gaoguin, S. *Computerspiele und lebenslanges Lernen. Eine Synthese von Gegensaetzen*. VS Verlag fuer Sozialwissenschaften — Springer Fachmedien Wiesbaden GmbH, 2010, 2010.

12. Gluecksmomente beim Computerspielen. `https://www.derwesten.de/spiele/gluecksmomente-beim-computerspielen-id913493.html`.

13. Havranek, M., Langer, N., Cheetham, M., and Jncke, L. Perspective and agency during video gaming influences spatial presence experience and brain activation patterns. *Behavioral and Brain Functions 8* (2012), 1–13.

14. Gaming for Health: Systematic Review and Meta-analysis of the Physical and Cognitive Effects of Active Computer Gaming in Older Adults. `https://academic.oup.com/ptj/article-abstract/doi/10.1093/ptj/pzx088/4097725?redirectedFrom=fulltext`.

15. Hellstroem, C., Nilsson, K. W., Leppert, J., and Aslund, C. Effects of adolescent online gaming time and motives on depressive, musculoskeletal, and psychosomatic symptoms. *Upsala Journal of Medical Sciences 120* (2015), 263–275.

16. Jeong, E. J., Kim, D. J., Lee, D. M., and Lee, H. R. A study of digital game addiction from aggression, loneliness and depression perspectives. *System Sciences (HICSS)* (2016), 3769–3780.

17. Das deutsche Jugendschutzsystem im Bereich der Video- und Computerspiele. `https://www.hans-bredow-institut.de/uploads/media/Publikationen/cms/media/b28a902bdb97d14f33417cbfb7459919ea205e6d.pdf`.

18. Kishino, F., Milgram, P., Takemura, H., and Utsumi, A. Augmented reality: a class of displays on the reality-virtuality continuum. *Proc. of SPIE 2351* (1995), 282–292.

19. Maercker, A., Hecker, T., and Heim, E. Personalisierte internet-psychotherapie-angebote fuer die posttraumatische belastungsstoerung. *Der Nervenarzt 86* (2015), 1333–1342.

20. Mandryk, R., Birk, M. V., Lobel, A., van Rooij, M., Granic, I., and Abeele, V. V. Games for the assessment and treatment of mental health. *Proc. of CHI PLAY '17 Extended Abstracts* (2017), 673–678.

21. Meehan, M., Insko, B., Whitton, M., and Brooks, F. P. Physiological measures of presence in stressful virtual environments. *Proc. of the 29th annual conference on Computer graphics and interactive techniques 21* (2002), 645–652.

22. Unraveling the enigma of Nintendos virtual boy, 20 Years later. `https://www.fastcompany.com/3050016/unraveling-the-enigma-of-nintendos-virtual-boy-20-years-`

23. Pertaub, D.-P., Slater, M., and Barker, C. An experiment on public speaking anxiety in response to three different types of virtual audience. *Presence: Teleoperators and Virtual Environments 11* (2002), 68–78.

24. Why Playing Video Games Can Actually Be Good for Your Health. `http://time.com/4051113/why-playing-video-games-can-actually-be-good-for-your-he`

25. 4 Reasons Video Games Are Good For Your Health (According To American Psychological Association). `https://www.forbes.com/sites/jordanshapiro/2013/11/27/4-reasons-video-games-are-good-for-your-health-according#301d9cb83a00`.

26. Robertson, B. Immersed in art. *Computer Graphics World 24* (2001).

27. Sallge, M. *Das Spiel: Muster und Metapher der Mediengesellschaft,*. VS Verlag fuer Sozialwissenschaften — Springer Fachmedien Wiesbaden GmbH, 2010.

28. Sega VR: Great idea of wishful thinking?
    `http://www.sega-16.com/2004/12/`
    `sega-vr-great-idea-or-wishful-thinking/`.

29. Shin, H. H., and Kim, K. M. Virtual reality for cognitive
    rehabilitation after brain injury: a systematic review.
    *Journal of Physical Therapy Science 27* (2015),
    2999–3002.

30. Simeonov, P., Hsiao, H., Dotson, B. W., and Ammons,
    D. E. Height effects in real and virtual environments.
    *Human Factors 47* (2005), 430–438.

31. Prognose zum Umsatz im Markt fuer Videogames
    weltweit von 2011 bis 2020 nach Segment (in
    Milliarden US-Dollar). `https://de.statista.com/`
    `statistik/daten/studie/160518/umfrage/`
    `prognostizierter-umsatz-in-der-weltweiten-videogames-branche/`.

32. Virtual reality (VR) headset installed base worldwide
    from 2014 to 2020 (in millions).
    `https://www.statista.com/statistics/619729/`
    `global-vr-headset-installed-base/`.

33. Virtual reality device unit shipments worldwide by
    vendor/brand in 2016 (in 1,000s)).
    `https://www.statista.com/statistics/705476/`
    `global-virtual-reality-device-shipments-by-vendor/`.

34. Virtual reality software and hardware market size
    worldwide from 2016 to 2020 (in billion U.S. dollars).
    `https://www.statista.com/statistics/528779/`
    `virtual-reality-market-size-worldwide/`.

35. Sturman, D. J., and Zeltzer, D. A survey of glove-based
    input. *IEEE Computer Graphics and Applications 14*
    (1994), 30–39.

36. Sutherland, I. The ultimate display. *Proc. of IFIP
    Congress* (1965), 506–508.

37. Brains reaction to virtual reality should prompt further
    study, suggests new research by UCLA neuroscientists.
    `http://newsroom.ucla.edu/releases/`
    `brains-reaction-to-virtual-reality-should-prompt-further-study-suggests`
    `-new-research-by-ucla-neuroscientists`.

38. Harmlose Videospiele machen gesund.
    `https://www.welt.de/wissenschaft/article155768/`
    `Harmlose-Videospiele-machen-gesund.html`.

39. Valves VR is seriously impressive. Its also got some
    issues. `https://kotaku.com/`
    `valves-vr-is-seriously-impressive-its-also-got-some-is-1689916512`.

40. Winerman, L. A virtual cure. *Monitor on Psychology 36*
    (2005), 87.

# YOUR KNOWLEDGE HAS VALUE

- We will publish your bachelor's and
  master's thesis, essays and papers

- Your own eBook and book -
  sold worldwide in all relevant shops

- Earn money with each sale

Upload your text at www.GRIN.com
and publish for free